Active Arithmetic and Algebra

Activities for Prealgebra and Beginning Algebra

Judy Jones
Valencia Community College

Brooks/Cole Publishing Company

I(T)P®   An International Thomson Publishing Company

Pacific Grove • Albany • Belmont • Bonn• Boston• Cincinnati
Detroit• Johannesburg • London • Madrid • Melbourne • Mexico City
New York • Paris • Singapore • Tokyo • Toronto • Washington

RWP

Assistant Editor: *Melissa Henderson*
Editorial Assistant: *Shelley Gesicki*
Marketing Team: *Sue Ewing and Jennifer Huber*
Marketing Assistant: *Debra Johnston*
Production Editor: *Mary Vezilich*
Cover Design: *Christine Garrigan*
Printing and Binding: *Mazer Corporation*

For more information, contact:

BROOKS/COLE PUBLISHING
511 Forest Lodge Road
Pacific Grove, CA 93950
USA

International Thomson Editores
Seneca 53
Col. Polanco
11560 México, D. F., México

International Thomson Publishing Europe
Berkshire House 168-173
High Holborn
London WC1V 7AA
England

International Thomson Publishing Japan
Hirakawacho Kyowa Building, 3F
2-2-1 Hirakawacho
Chiyoda-ku, Tokyo 102
Japan

Thomas Nelson Australia
102 Dodds Street
South Melbourne, 3205
Victoria, Australia

International Thomson Publishing Asia
60 Albert Street
#15-01 Albert Complex
Singapore 189969

Nelson Canada
1120 Birchmount Road
Scarborough, Ontario
Canada M1K 5G4

International Thomson Publishing GmbH
Königswinterer Strasse
53227 Bonn 418
Germany

Printed in the United States of America

5   4   3   2   1

ISBN 0-534-36771-2

TABLE OF CONTENTS

Dear Student,

Listening to the teacher! Reading the textbook! Working problems from the book! Taking tests!

Do these bring back memories of your math classes? If so, you have not really "experienced" mathematics.

This book contains activities that will let you get more involved with your study of math. You will work with partners or teams to discover mathematical concepts or to verify a mathematics formula. You will collect real data, draw graphs, measure volume, or work puzzles.

Have you ever asked your mathematics instructor when you will ever use math? Hopefully, some of these activities will show you some areas that you could use math effectively <u>outside</u> of the classroom. Did you know that ratios and proportions are important to artists? That you can make a scale drawing of your room to help calculate the area of the walls before buying paint?

Arithmetic and algebra are useful tools in all aspects of your life. The activities in this book provide some examples in how they may be used.

Judy Jones

BAR GRAPHS AND LINE GRAPHS

Data can be collected and displayed in many ways. Among these are tables, bar graphs, line graphs, and circle graphs. In this activity, you will interpret both bar and line graphs and will also create tables and graphs from real data.

PART I
INTERPRETING TABLES AND GRAPHS

In a city in the southwestern part of the United States, the newspaper reports the length of time a normal individual may be in the sun without using any sunscreen. The data for one particular day is shown below as a table, and on the next page as a bar graph and a line graph.

TABLE

Maximum Amount of Time to Spend in the Sun

Time	Maximum Minutes
9:00 AM	34
10 :00 AM	20
11:00 AM	15
12:00 PM	13
1:00 PM	14
2:00 PM	18
3:00 PM	32
4:00 PM	60

BAR GRAPH

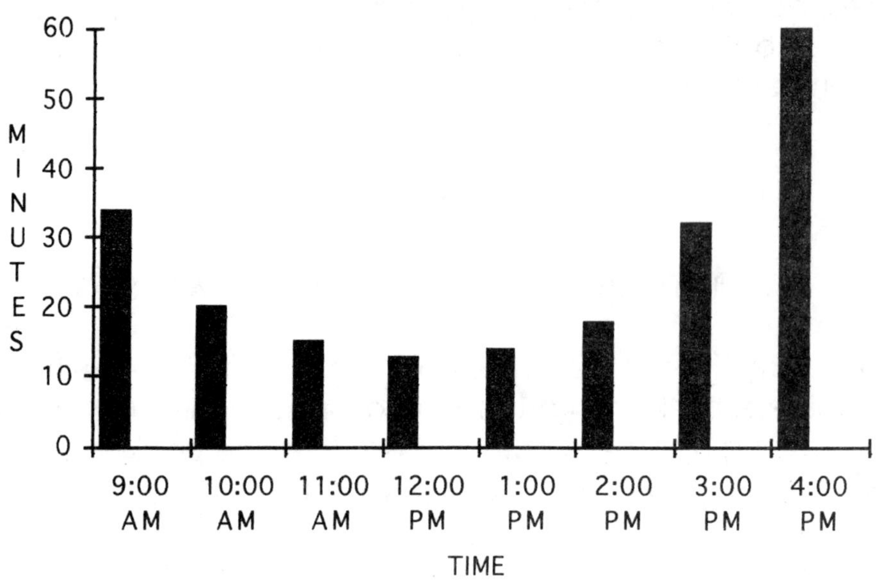

Maximum Amount of Time to Spend in the Sun

LINE GRAPH

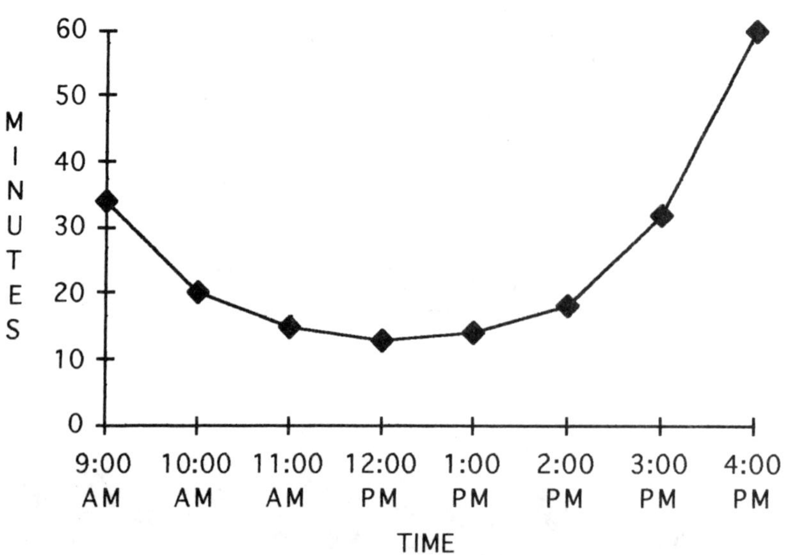

Maximum Amount of Time to Spend in the Sun

Answer the following questions using the table and the two graphs above. Write answers in complete sentences where appropriate.

1. How long could you spend in the sun before burning if you were outside at 10:00 AM?

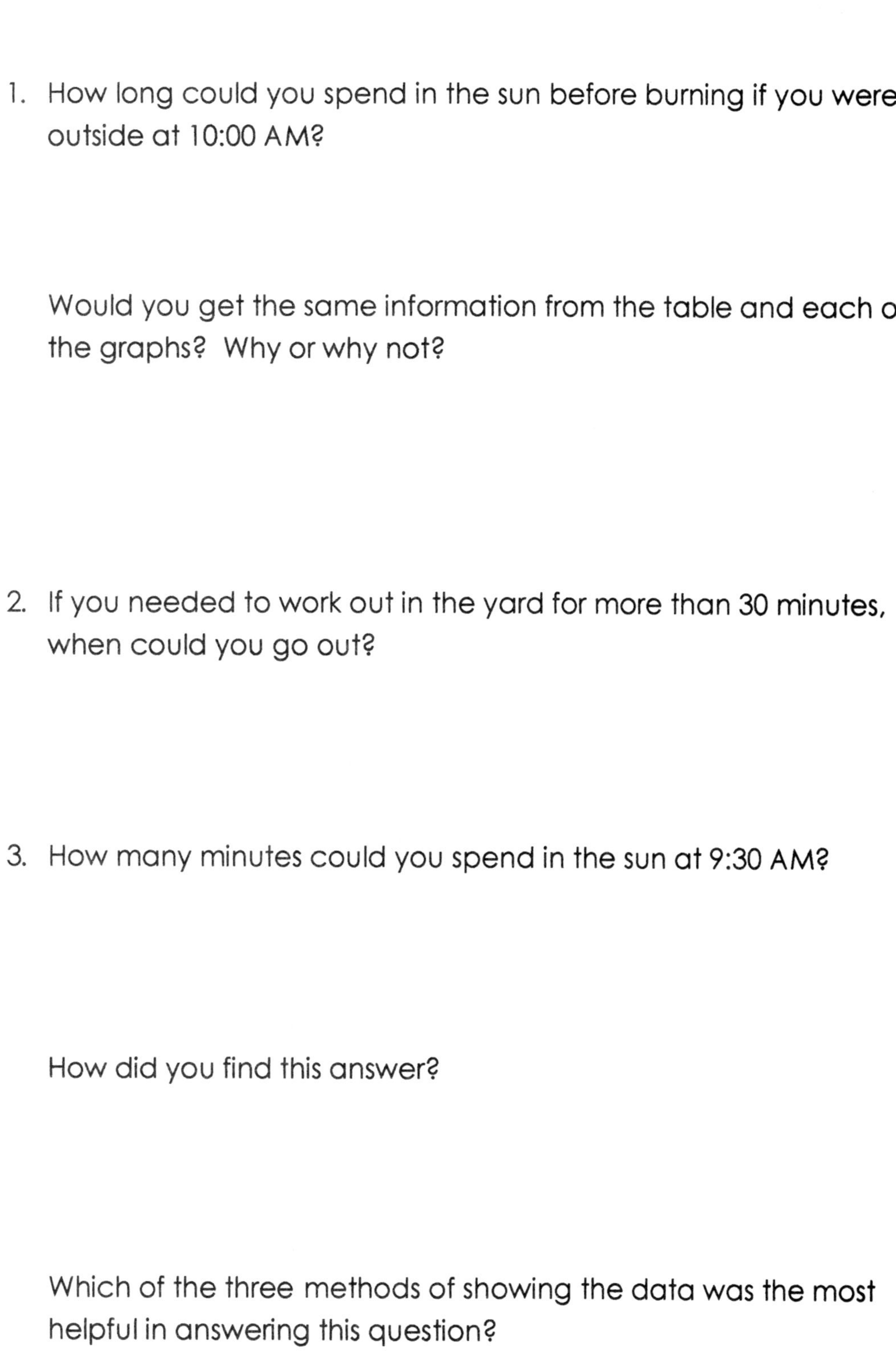

    Would you get the same information from the table and each of the graphs? Why or why not?

2. If you needed to work out in the yard for more than 30 minutes, when could you go out?

3. How many minutes could you spend in the sun at 9:30 AM?

    How did you find this answer?

    Which of the three methods of showing the data was the most helpful in answering this question?

4. Could you decide how long to stay outside before 9:00 AM?
Why or why not?

PART II
CONSTRUCTING TABLES AND GRAPHS FROM REAL DATA

1. As a class, find out how many students are taking one
class, two classes, three classes, etc. this semester. Fill in the
data in the table below.

Number of Classes	Number of Students

2. Work with a group of 3-4 students and decide how to graph this data. The following questions should be discussed.

   What will you use as a title for your graphs?

   What data will you plot along the horizontal axis?

   What will you plot along the vertical axis?

   How far apart will you place the bars in your bar graph?

   How much will each unit on your graph paper represent horizontally for your line graph?

   How much will each unit on your graph paper represent vertically?

3. Each member of the group should draw both a bar graph and a line graph using the data in the chart above on graph paper. Be sure to label the axes, show the scale, and give each graph a title.

4. How many students are taking 2 classes?

5. Would it make sense to ask how many students are taking 2.5 classes? Why or why not?

PART III
CREATING TABLES AND GRAPHS FROM POPULATION DATA AND MAKING PREDICTIONS

Each group should get a set of population reports from your instructor. These reports may give information about total population, population by sex, population by age, and/or population by race and ethnic origin. You have reports for several years. <u>Each member</u> of the team should choose <u>a different type</u> of measure, such as total population, and make a table, a bar graph and a line graph for that measure over a period of years. Be sure to decide what will be graphed along the horizontal axis and what will be graphed along the vertical axis. When each team member has completed his/her table and graphs, compare the results and answer the following questions.

1. Do each of the graphs have the same general shape?

    If they do, what does this mean?

    If the shapes are different, why do you think this might be?

2. Is the population of all of the groups that you studied increasing?

    Does one group seem to be increasing more rapidly? Which group? Why might this particular group be growing faster than the others?

3. You are a member of the County Commission, and you have to make plans for the future growth. Working with your group, use your population data to make your recommendations. Consider issues such as public transportation, childcare, retirement homes, schools, hospitals, police and fire protection, etc.

   Present your recommendations to the class using your data and graphs.

ADDITION AND SUBTRACTION WITH INTEGERS

Using visual models will help you understand better how to add and subtract using integers. Integers include the set of whole numbers and their opposites. Included in this lab on page 7 is a set of positive and negative boxes. Check with your instructor to see whether you need to cut this sheet apart before your lab class.

Part 1
Understanding the Boxes

a. Each of the boxes with a $\boxed{+}$ in it represents (+ 1), and each of the boxes with a $\boxed{-}$ in it represents (− 1). To represent − 3, we will use 3 of the $\boxed{-}$ boxes; to represent a + 2, we will use 2 of the $\boxed{+}$ boxes.

b. The only integer that is not represented with the boxes is 0. We will need to use 0 during this lab activity, so we will represent 0 using an equal number of $\boxed{+}$ and $\boxed{-}$ boxes. For example, $\boxed{+}\boxed{-}$ represents 0. Zero can also be represented by any multiple of positive and negative boxes. $\boxed{+}\boxed{-}\boxed{+}\boxed{-}\boxed{+}\boxed{-}$ can also represent 0.

c. For practice, use your boxes to represent the integer, − 6. (Make a sketch below.)

d. Add additional boxes to the boxes used in part c to form the number 0. What type of boxes did you use?

How many boxes were required?

Complete the following − 6 + (+6) =

Part 2
Addition of Integers

a. You have added positive integers since elementary school, so write the solution to the following problems without using the boxes.

$$(+2) + (+5) = \qquad\qquad (+6) + (+8) =$$

Notice that there is a + sign belonging to the number indicating that the number is a positive number and that there is also a + indicating addition. We usually don't always write the + sign in front of positive numbers, but remember that it is there.

Write a rule for adding two positive integers.

b. Use the boxes to model the addition of two negative numbers.

$$(-2) + (-5) =$$

$$(-2) \quad + \quad (-5) \quad =$$

Sketch the result from this problem.

$$(-6) + (-8) =$$

Write a rule for adding two negative integers using the result above.

Without using the boxes find: $(-22) + (-7) + (-4) + (-10) =$

c. Use the boxes to model the addition of a negative number and a positive number.

$(-2) + (+5) =$ $\boxed{-}\boxed{-}$ $+$ $\boxed{+}\boxed{+}$ $\boxed{+}\boxed{+}$ $\boxed{+}$ $=$?

Pair up as many $\boxed{+}$ and $\boxed{-}$ boxes as possible. What does each pair equal? 0 How many boxes do you have left over?   What integer do they represent?

$(-2) + (+5) =$ _________

Use your boxes to model the following problems and then write the integer that is the solution to the problem.

$(-6) + (+4) =$ (sketch the boxes here)

$(-6) + (+4) =$ _____________

$5 + (-6) =$ (sketch the boxes here)

$5 + (-6) =$ _____________

$4 + (-2) =$ (sketch the boxes here)

$4 + (-2) =$ _____________

Write a rule for adding a negative integer and a positive integer using the results above. (Is the solution always negative or always positive?)

Try the following problem without using the boxes. Verify your answer by using the boxes.

$$6 + (-4) + (-5) + 7 =$$

Part 3
Subtraction with Integers

a. We want to use the "take away" concept for subtraction in our models for subtraction of integers. Let's model $(+6) - (+4) = ?$

$$\boxed{+}\,\boxed{+}\,\boxed{+}\,\boxed{+}\,\boxed{+}\,\boxed{+} \quad \text{take away } 4\ \boxed{+} \quad =$$

$$\boxed{+}\,\boxed{+}\,\boxed{\boxed{+}\,\boxed{+}\,\boxed{+}\,\boxed{+}} = \boxed{+}\,\boxed{+}$$

$$(+6) - (+4) = \underline{+2}$$

b. Model the solution to the following problem:

$(-6) - (-2) =$ (Sketch the boxes and show the "taking away" as we did in part a.

$$(-6) - (-2) = \underline{\qquad\qquad}$$

c. To model subtraction with numbers with opposite signs, we need to use the concept of 0 as we discussed at the beginning of the activity.   Find the answer to the following problem using the boxes.

$$(-6) - (+4) =$$

We can represent the -6 as before by using 6 $\boxed{-}$ boxes, but we can't take away 4 $\boxed{+}$ boxes because we don't have any. Since adding 0 to any number will not change the number, we will add 0 in the form of $\boxed{+}\boxed{-}\boxed{+}\boxed{-}\boxed{+}\boxed{-}\boxed{+}\boxed{-}$. Now we have the following set of boxes:

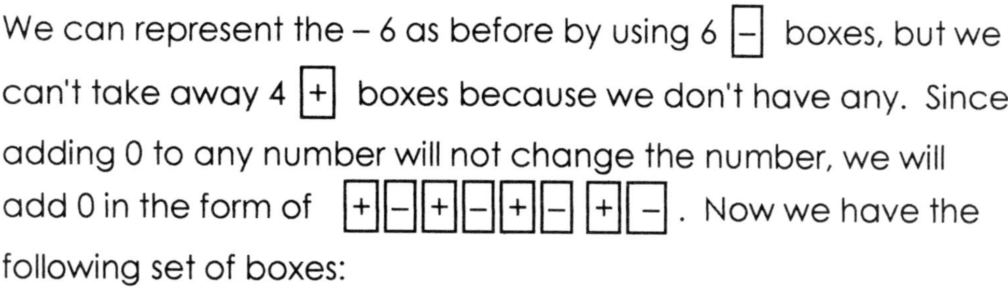

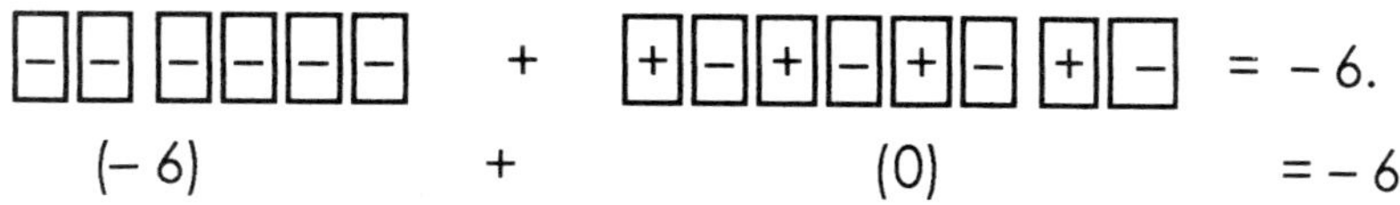

Now if we remove 4 of the $\boxed{+}$ boxes, you will have 10 $\boxed{-}$ boxes left.

Therefore, $(-6) - (+4) = -10$

d. Use the same concept as we did in part c to find the solution to:

$6 - (-3) =$ (Sketch the boxes and show the "taking away")
How many pairs of $\boxed{+}$ and $\boxed{-}$ boxes should you add?

$6 - (-3) =$ ________________

e. Now try the problem: $(-5) - (-8) = ?$
Use the boxes first and then write the solution.
$(-5) - (-8) =$ ________________

f. Before you try to write a rule for subtraction, compare the
    answers to the following problems. The second problem in each
    set follows the addition rules from part 2.

1. $(+6) - (+4) =$

    $(+6) + (-4) =$

2. $(-6) - (-2) =$

    $(-6) + (+2) =$

3. $(-6) - (+4) =$

    $(-6) + (-4) =$

4. $(-5) - (-8) =$

    $(-5) + (+8) =$

If you have done your calculations correctly, you should see that
the answers in each set are the same. Write a rule for finding the
solution to a subtraction problem that includes changing the
subtraction to an addition.

Find the solution to each of the following subtraction problems
by first writing a new problem that is an addition problem.

$-18 - (+9) =$ _________________________________ = _______________
                      (addition)

$13 - (-5) \quad =$ _________________________________ = _______________
                      (addition)

$-20 - (-8) \quad =$ _________________________________ = _______________
                      (addition)

Part 4

Make a copy of this page to use in this lab activity. Cut out the boxes. Check with your instructor to find out whether you should have them prepared prior to doing the activity.

+	+	+	+	+	+	+	+	+	+
+	+	+	+	+	+	+	+	+	+
−	−	−	−	−	−	−	−	−	−
−	−	−	−	−	−	−	−	−	−

A POLYNOMIAL PUZZLE

Algebraic expressions can be manipulated through addition, subtraction, multiplication and division. We can also evaluate expressions by substituting real numbers for the variables. Translations from English to algebra or vice versa are also a necessary skill to be successful in prealgebra and algebra courses. The polynomial puzzle requires many of the techniques described to be able to complete it successfully. Two of the entries are already completed for you. If a number is negative, include the sign within the same box with the number. If you are using addition, subtraction, or parenthesis signs, put them in separate boxes. You may use the bottom of this page to work on the clues.

TURN THIS SHEET OVER TO FIND THE PUZZLE AND THE CLUES.

ACROSS	DOWN

ACROSS

2. $3x(4x + 5)$
4. $5y + 3 - 2y + 7 - 6x$
5. $(-9)(4)$
8. $10x^2 - 14y + 4 - 6y - 9x^2 - 2$
10. The difference between z and 12
11. The sum of 5 and t
14. $-7(-7x + 2)$
15. Evaluate $x^2 - 20x + 56$ for $x = -45$
16. $4(3r - t - 78)$
18. $11(p - 3q + 1)$
20. The product of the numbers d, o, n, and e

DOWN

1. $10x - 35x + 4 + 18$
3. $2(x - 3y + z)$
6. $7 - t + 3m - 4m + 42 - 6t$
7. 4 more than the product of y and 6
10. $4m + 16 - 2m$
12. $15 + 10v + 30r + 19v - 27r - 13$
14. Evaluate $2m^3 + 8m + 65$ for $x = 15$
17. The area of a rectangle with sides of 63 and 51
19. The difference between q and the number n

THE ALGEBRAIC MATCH-UP

It is important that you learn to read and speak using appropriate algebraic vocabulary. This activity will give you practice in translating English phrases to algebraic expressions and algebraic expressions to English phrases. Your teacher will provide each group of students with a set of Algebraic Match-up cards. The instructions for playing the game follow.

1. Shuffle and lay out all of the cards face down in 4 rows of 6 cards each. Be sure that each student has paper and a pencil. The goal of the game is to match as many pairs of cards as possible. A pair will have an English phrase and the corresponding algebraic expression.

2. Students should play in order of their birthday. Use the day of the month that each student was born. If there are two students born on the same day, then use the month also. The student with the earliest birthday in the year will begin, followed by the next earliest, etc.

3. The first player should turn over two cards and read each of the cards using the English or algebraic symbols showing on the card first. Then each card should be translated into the opposite language. For example, if one of the cards turned over has the English phrase, 5 more than a number, the student should read the phrase as written and then should translate the phrase to algebraic symbols by saying, $5 + x$, $n + 5$, or any other equivalent expression. If an algebraic expression such as $2p$ appears, the expression should be read as written and then translated into an equivalent English phrase such as the product of 2 and a number, twice a number, or other equivalent phrase.

4. If anyone in the group disagrees with any of the translations, the group should discuss the card and come to a consensus. If the two cards match cards, the student takes both cards and

continues by turning two more cards over. If they do not match, both cards should be turned back over and placed in their original places. (It will be helpful to remember where each of the cards is placed.)

5. The next student takes a turn by turning over any two cards, reading the phrases and translating.   Play progresses until all of the cards have been paired and removed from the table.

6. There are two sets of cards. If you finish the game with the first set, ask your instructor for the second set of cards. The rules are the same for the second set of cards, but the phrases are more difficult.

UNDERSTANDING FRACTIONS

Fractions are a part of our everyday lives. We watch our fuel gauge on the car move from full to 1/2 to 1/4 to empty. We use notebook paper that measures $8\frac{1}{2}$ inches by 11 inches. We make cookies using 1 cup of butter and $2\frac{1}{2}$ cups of flour.

To visualize fractions, a geometric shape is often used. For example: the rectangle below has $\frac{2}{6}$ or $\frac{1}{3}$ of the area shaded.

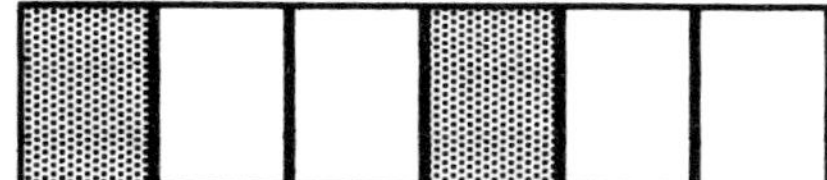

You will work in groups of 2-3 students to shade areas of equal size and also to write a fraction that describes the shaded part of geometric shape.

Part 1

Use dot paper or a geoboard to represent each of the following shaded areas. Draw each square with lengths of 6 units on each side. Find the fractional part of the square that is shaded. Be sure to reduce your fraction to lowest terms. Describe the method that you used and/or show the calculations that verify your answer under each diagram below.

Each member of your team should write the answer to at least 2 of the problems. Be sure that each member agrees with the answers to all of the problems.

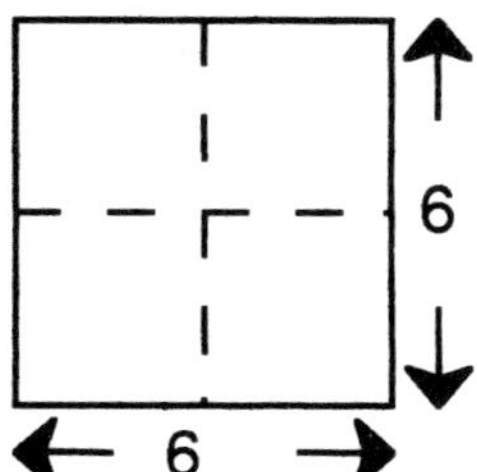

a.

b.

c.

d.

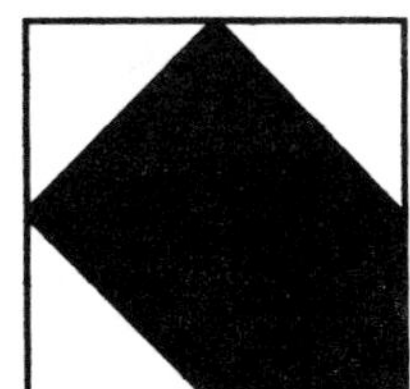

e.

f.

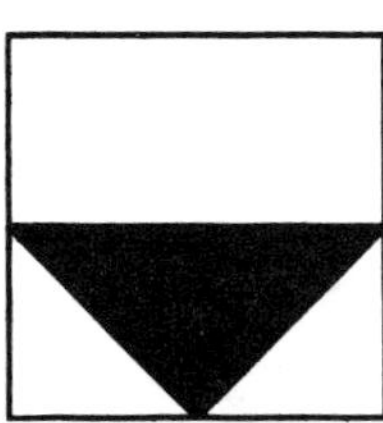

g.

h.

i.

j.

Part 2

Rectangles or squares can be divided into fractional parts such as $\frac{1}{2}$ fairly easily. Some triangles can also be divided into fractional parts easily.

a. Use the dot paper to draw the isosceles triangle below and divide the triangle so that you have two equal areas.

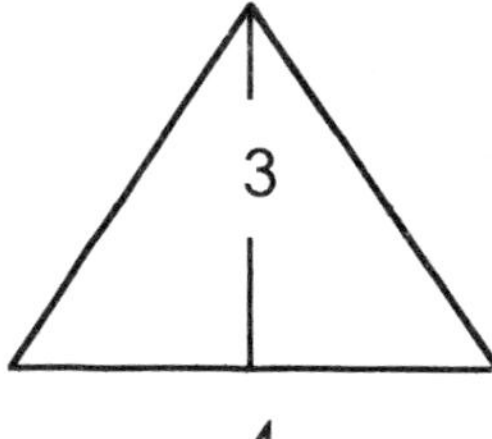

Find the area of the large triangle and also the area of the two smaller triangles. Hint: To find the area of a triangle, use the formula $A = \frac{1}{2}bh$ where b is the base of the triangle and h is the height of a perpendicular line drawn from the base. Is the area of one of the smaller triangles one-half the area of the large triangle?

b. Use the dot paper to draw an obtuse scalene triangle similar to the one below. Let the distance between each dot represent one unit of length. Be sure your triangle has a base that is 8 units long and the height is 4 units.

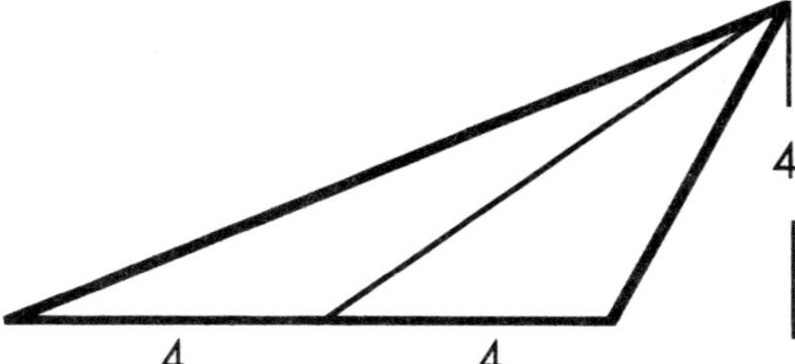

The triangle is divided into two parts. Show that the two areas are equal.

c. Draw the triangle shown below on the dot paper using the distance between 2 dots to represent 1 unit of length.

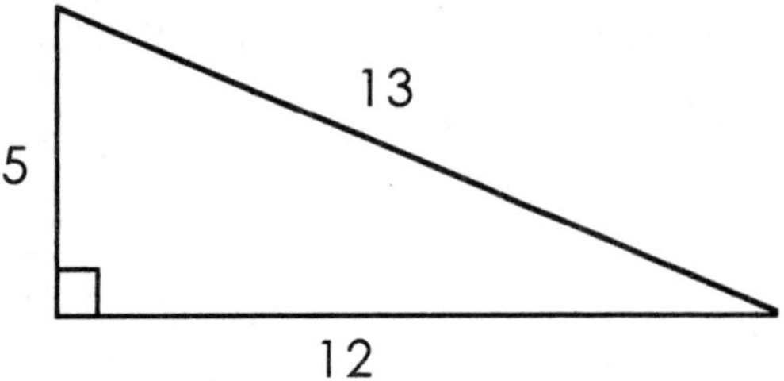

In the two previous triangles, you were able to divide them into halves. How did you divide the base of the triangle to form a smaller triangle with one-half of the area of the large triangle?

Use this idea to decide how to draw lines separating the triangle on your dot paper into thirds (three smaller triangles each with an area of $\frac{1}{3}$ the area of the large triangle.) <u>Show</u> that each area is $\frac{1}{3}$ of the area of the large triangle.

Calculations:

THE PREALGEBRA PIZZA PARTY

Fractions, decimals and percents all represent part of a whole. Sometimes it is easier to use one form than another. For example, if you have divided a rope into three equal pieces, it is easier to describe one of the pieces as $\frac{1}{3}$ of the rope than as the decimal equivalent, $0.\overline{3}$. We describe the sales tax in Florida as 6 %, not as $\frac{3}{50}$. Complete all sections of the activity below.

Angelica, Bryan and Cheryl were studying for a prealgebra test when they decided to order pizza. Because they were very hungry and expected to spend several more hours studying, they ordered 4 medium pizzas. When the pizzas arrived, they discovered that two of the pizzas were cut in half, one pizza was cut into three equal pieces and the fourth one was cut into four equal parts. Cut the models of the pizzas into the appropriate number of pieces and/or complete the sketch of the pizzas below showing how each was cut as your instructor directs.

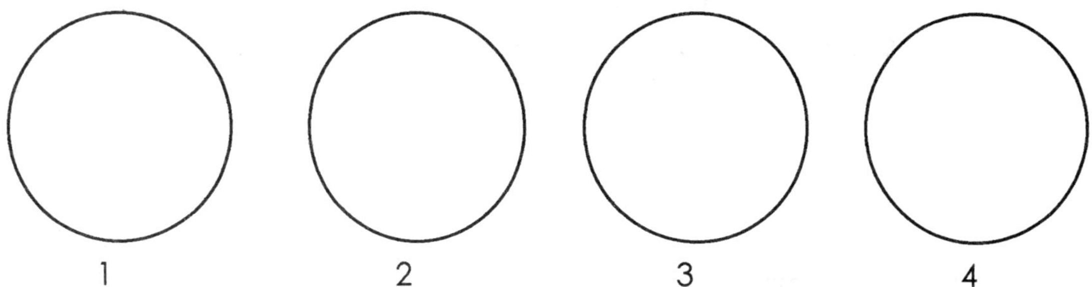

If each of the students took one piece from the first two pizzas (each cut in halves), one from the third (cut in thirds) and one from the fourth pizza (cut in fourths), answer the following questions. <u>Show any calculations that you use, and write your answers using complete sentences</u>. (You may find it helpful to label the pieces of pizza with the name of the person that took that piece.)

1. Use a fraction to express the <u>total</u> amount of pizza that each student took.

2. a. After each of the students finished eating his/her pizza, how many <u>pieces</u> of pizza were left?

    b. Were all of the pieces equal in size?

    c. Write the <u>total</u> amount of pizza that was left as a fraction.

    d. If the leftover pizza was <u>equally divided</u> among the three students, what fractional part of a pizza would each one get?

        How could the leftover pizza be equally divided? Be sure to describe in complete sentences.

3. a. If each pizza cost $6.95, what was the value of the pizza that each student <u>ate</u>? (Round off to the nearest cent.)

    b. What was the value of the <u>total</u> amount of leftover pizza?

4. In the town where Angelica, Bryan, and Cheryl live, there is a 12% delivery charge for each pizza.

   a. What was the total delivery charge for the four pizzas?

   b. How much would each student have to pay for the delivery charge if it was divided equally?

   c. What was the total cost of the pizza including the pizza each ate, the leftover pizza, and the delivery charge for <u>each</u> of the students?

5. a. Use the result from Question 1 and write the amount of pizza that each student ate in decimal form. (Round off if necessary to three decimal places.)

   b. Write the total amount of pizza that was left as a decimal.

   c. Write the amount of leftover pizza each student received as a decimal.

d. Recalculate the total cost of the pizza each student ate using the decimal form found in 5a.

e. Recalculate the total cost of the pizza that was left over using the decimal form from 5b.

f. Use the results from parts 4 and 5 to find the total cost of the pizza and the delivery charge for each student.

Does the answer agree with the result from 4c?

Why or why not?

6. Which type of numbers (fractions or decimals) did you prefer to work with?

Why? Be specific.

EQUATION TRIANGLES

This activity will combine solving equations and solving a geometric puzzle. The skills that you use to simplify algebraic expressions will be needed as well as the properties of equality.

You will work with a group of two or three students. Each group will receive a set of triangles with equations and answers. Each equation shown on the set of triangles must be solved and matching the equations with the solutions must make a geometric shape. (There are some extra answers and also some equations that do not have solutions listed on the triangles.) Your instructor will tell you whether you need to turn in individual worksheets or whether you may turn in one group report. Be sure to attach any work that you did in solving the equations to the solutions.

Equation	Solution
A-1	
A-2	
B-1	
C-1	
C-2	
D-1	
E-1	
F-1	
F-2	
G-1	
G-2	

EQUATION	SOLUTION
H-1	
I-1	
J-1	
K-1	
K-2	
L-1	
L-2	

Draw a diagram below showing the way that you put the triangles together. Be sure to label each one with A – L so that your instructor can see the solution to your puzzle.

THE BOUNCING BALL

Real world data can be collected using technology. You will work with a partner to measure the initial height of a ball and the height of the first bounce after dropping the ball using a TI graphing calculator and a CBR™. The Calculator-Based Ranger is a portable sonic motion detector. The data collected will be entered in a table and plotted as points on a graph. You will draw the best "eyeball fit" line, and will use the line to predict the initial heights and the height of the first bounce.

Get a ball, graphing calculator and CBR™. (Your instructor may have stations with the equipment already set up.) Follow the steps listed below to get the four different measurements for dropping the ball. (Be sure to drop the ball from a different height each time.)

1. Connect the graphing calculator to the CBR™ using the cable provided. Be sure that both connections are tight.

1. Turn on the calculator and push the button marked PRGM. Choose the RANGER program by highlighting the number in front of the name and pushing ENTER or by typing the number in front. PrgmRANGER will appear on the calculator screen. Push ENTER. Follow the instructions on the next screen.

1. The MAIN MENU will list 6 options. Choose the option APPLICATIONS. The next screen will give you the option of measuring in meters or feet. Your instructor will tell you which units to use. Choose the appropriate unit by highlighting and pushing ENTER or by typing the correct number.

1. The next screen will list the applications. Choose BALL BOUNCE.

1. The next screen will look like this.

6. Push ENTER. A new screen will appear that looks like this.

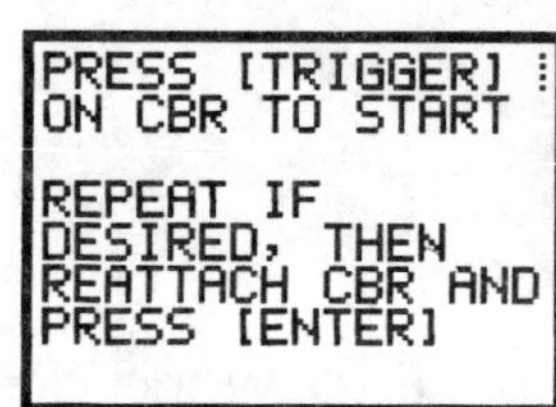

7. Do not push ENTER yet. Follow the directions in the next paragraph.

8. Because the CBR™ cannot measure distances closer than 0.5 meters (about 18 inches), you will have to keep the CBR™ at least 0.5 meters above the ball when you begin your data collection. If you do not know how long 18 inches is, use your ruler to help. One of the partners should hold the CBR™ about 5-6 feet above the floor with the sensor pointing toward the floor. The other partner should hold the ball about 18 inches below the CBR™ with arms extended. When you both are in position, practice one or two bounces of the ball. The CBR™ should remain in position and the ball should bounce 2-3 times directly below the CBR™ sensor. When you have the technique for bouncing set, the partner with the CBR™ should push the trigger button and as soon as the sound begins, the partner should drop the ball. When the sound ceases, push ENTER. (If you have disconnected the CBR™ from the calculator during the ball bounce, reconnect the two before pressing ENTER.) You will see a message on the screen while the data is being transferred to the calculator screen.

9. You will get a screen that will look similar to the one below depending on the number of bounces that was recorded by the CBR™.

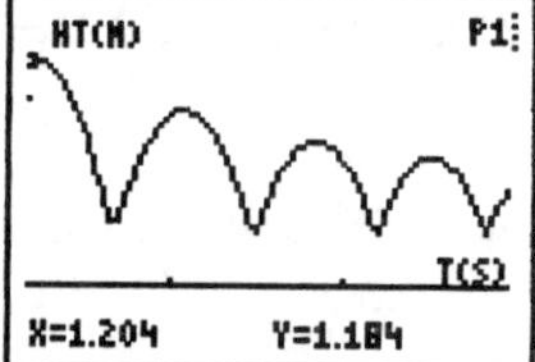

10. Use the arrow keys to move the cursor to the right. Get the largest reading in the first peak for the initial height of the ball. Then move to the top of the next peak and get that reading. That will be the height of the first bounce. Record each of the values in the table below. Press ENTER when you have recorded the values and choose REPEAT SAMPLE. You will return to the first screen in Step 5. Repeat the collection for a total of 4 points. Be sure to drop the ball from a different height each time. Record all of the values in the table.

Initial Height of Ball	Height of First Bounce

11. Use graph paper to plot the 4 points. Start each axis at 0 and use your data to decide on the scale necessary. Round off the measurements to the nearest hundredth when you plot the points. Draw the "best" straight line that you can through your 4 points. If your points do not seem to fall in a straight line, repeat one or more measurements until you get 4 points that resemble a line. Connect the four points and extend the line in both directions as far as your graph will allow.

12. Use your graph to estimate what the height of a bounce should be if the <u>initial height</u> of the ball was 3 feet (0.9 meters).

13. Use your graph to estimate how high you should hold the ball to get the height of the <u>first bounce</u> to be 3 feet (0.9 meters).

14. If other pairs of students used a different type of ball than you did, compare your data in the tables and the graphs of the lines. How are they different? Be specific.

How are they the same? Be specific.

What difference(s) in the balls would account for the difference(s) in the graphs? Be specific.

PAINTERS' PAY

One of the goals of taking a mathematics course is to be able to apply the methods learned to a real life situation. You will use **tables** and **graphs** to describe the relationship between the earnings of a painter and the painter's helper.

Two painters are scheduled to work on Monday at an apartment complex. One of the painters, Max, has only been working as a painter for a short time and earns $7 an hour. His partner, Carmen, has several years of experience and earns $11.50 an hour. Max begins work at 8:00 A.M. , but Carmen has not arrived. When Carmen arrives at 10 A.M., she explains that she was delayed by an accident on the interstate. Both partners work until 5 P.M.

1. Max realizes that even though he earns less money each hour he might earn more money than Carmen today because she didn't work as many hours. Fill out the following table that describes their work day and their earnings. (They get paid for their lunch break.)

EARNINGS TABLE

Time	Total Hours worked Max	Total Earnings Max	Total Hours worked Carmen	Total Earnings Carmen
8 A.M.	0	0	0	0
9 A.M.	1	$7	0	0
10 A.M.			0	0
11 A.M.				
12 noon				
1 P.M.				
2 P.M.				
3 P.M.				
4 P.M.				
5 P.M.				

2.   Look carefully at the values for "total hours worked" and "total amount" earned for Max.

a. What relationship is there between these two numbers? (Use a complete sentence.)

b. Does this relationship hold for all entries in Max's columns?

c. Write an <u>algebraic</u> expression for the amount of money Max earned in m hours.

3.   Look at the total hours worked and "total amount earned" for Carmen.
a. What is the relationship between these two numbers?

b. Is the relationship the same as you found for Max? Why or why not?

c. Write an algebraic expression for the amount of money Carmen earns in c hours.

4.   a. When Max has worked 4 hours, how many hours has Carmen worked?

b. When Max has worked 6 hours, how many hours has Carmen worked?

c. Describe <u>in words</u> how long Carmen has worked when Max has worked for "m" hours.

d. If Max works m hours, write an <u>algebraic expression</u> for the number of hours that Carmen has worked using m hours. (Notice that in 3c, we used the variable c to represent the number of hours that Carmen worked.   In algebra, it is often easier to solve problems using only one variable. Instead of using both m and c , we will use only m.)

e. Using the new expression for the number of hours that Carmen worked (from 4d), write an expression for the total amount of money that <u>Carmen </u>earns if Max works <u>m</u> hours.

5.   a. When Max and Carmen both quit working at 5 P.M., who earns more money?

b. Use your table to estimate the time of the day when Max has still earned more money than Carmen.

6.   A graph can also show the information that is listed in your table. You will draw two different graphs on one set of axes so that you can compare Max's and Carmen's earnings.

a. Before you draw the axes on your graph paper, consider whether you should use any negative values. Should you show negative hours or negative earnings? Why or why not?

b. Use the horizontal axis to represent the number of hours worked. How many hours did Max work? How many hours did Carmen work? Decide what your scale should be using these values. Use the vertical axis to represent earnings starting with $0. Decide on what scale you should use by looking at the values of the earnings on your table.

c. Use the hours worked as your first coordinate and the total earnings as your second coordinate. For example, since Max had not worked any hours at 8 a.m., plot the point (0,0) for the first point on the graph for the Max. Graph each of the points for Max and then connect the points.

d. Repeat the graphing procedure for Carmen, but use a different color. Connect the points for Carmen.

e. Use your graph to estimate the number of hours that Max works and still earns more money than Carmen.

How did you decide?

f. Use your graph to estimate when Max and Carmen have earned the same pay.

7.   You can also find the <u>time of the day</u> when the earnings are equal by using an equation.

a. Set the expression for Max's earnings from 2c equal to the expression for Carmen's earnings from 4e and solve for m. Write the answer in terms of hours and minutes.

b. Since m represents the total hours that Max worked, find the time of day when the earnings are equal. Does the result agree with the graph? Why or why not?

DESCRIBING A LINEAR RELATIONSHIP

Real world data often can be described with a linear equation. You will work with a partner to the length of your forearm and your height in this activity. The data collected will be entered in a table and plotted as points on a graph. When the points are connected, a line will be the result. You will write the equation that describes a linear relationship between the two points and then use the equation to predict height or length of forearm.

1. Work with a partner whose height is several inches different from yours. Get a tape measure or yardstick and measure the height without shoes and the length of the right forearm (from elbow to tip of middle finger) of each of the partners. Placing your elbow against a book or a wall and laying it on the tape measure or yardstick will help give consistent measurements. Record each measurement in the table below.

Name of partner	Length of forearm	Height

2. Write each of the sets of measurements as an ordered pair using the length of the forearm as the first coordinate and the height as the second coordinate.

3. Use a full sheet of graph paper to plot your two points. Be sure to label each axis and to show the scale. You must start your scale

with 0 for both axes. Draw a line through your two points. Be sure to extend the line in both directions as far as your graph will allow.

4. A straight line has a unique characteristic. As you move from one point to the next the ratio of the number of units moved vertically to the number of units moved horizontally is a constant. Calculate this ratio for your line by completing the calculations below. Round off your answer to the nearest tenth.

$$\frac{\text{(your height } - \text{ partner's height)}}{\text{(your armlength } - \text{ partner's armlength)}}$$

5. The ratio that you calculated in part 3 is an important part of the equation that can describe your line. If we use A to represent the first coordinate and H to represent the second coordinate, we can begin to write the equation by using the ratio calculated above as the coefficient of A. Fill in your ratio in the equation below.

H = ______ A

6. Use the equation you wrote in part 5 to fill in the table below. Then plot the points on the same graph used earlier and draw a line through the points.

A	H
0	
10	
20	

What do you notice about the two lines that you have drawn?

How are they the same?

How are they different?

7. The first line that you drew is above your second line.

How many units above is it?

How did you find the change?

8. To complete the equation that describes the first line, we need
    to add the number of units that you found in part 6 to the
    equation. Complete the equation below by using the ratio from
    part 4 as the coefficient of A and then add number of units from
    part 6.

    $H =$ ___________ $A +$ ____________

9. Complete the table below using your new equation.

A	H
0	
10	
20	

10. Plot the points from the table and draw a line through the 3
    points extending it as before.

    Compare this line with the original line that you drew connecting
    the measurements of forearm and height. Be a specific as
    possible.

11. Get the forearm lengths from another set of students. Use your
    equation from part 8 to calculate each student's height. (Show
    your work below.)

    How closely do your results agree with the students' actual
    heights?

WRITING PERCENT PROBLEMS

The newspapers and magazines often use percent in articles and ads. In this activity you will develop application problems directly from current newspapers or magazines. You will share your problems with other members of your team and solve each problem. Your instructor will let you know if you need to bring in newspapers and magazines or if they will be provided.

WRITING THE PROBLEM

Write the problem(s) that you developed here. Be sure to give the source including title, date and page number for each of your problems.

SOLVING YOUR PROBLEM

Write a complete solution to your problem here. (Do not share the solution with your team members until everyone has a chance to solve your problem himself or herself.)

SOLVING THE TEAM PROBLEMS

Solve each of the problems developed by the other team members here.
After all of you have completed the problems, compare answers and
make corrections if necessary.   Be sure to identify each problem with the
team member's name. Use extra paper if necessary.

MATHEMATICS OF DRAWING THE HUMAN FIGURE

Ralph Larmann at Southern Arkansas University describes the standard proportions of the human body from the perspective of the artist at his web site on the Internet (http://www.saumag.edu/art/figure-drawing/body.html). His basis of measurement is the length of the head. Using a head as the measurement, most people fall between 6 heads and 8 heads tall with an average of 7 head tall. The shoulders are 3 heads wide. The distance from the wrist to the end of the fingers is one head. The distance from the elbow to the end of the fingers is 2 heads. You will work with a partner to see if you have the same proportions that an artist uses when drawing a human figure.

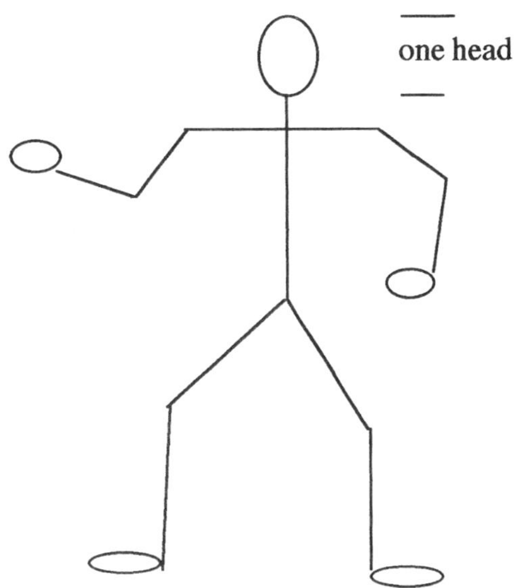

1. To measure the head, overall height, shoulder width, elbow to finger length and wrist to finger length, we will use a pencil as a measuring stick. Have your partner move 7-8 feet away from you. Hold a pencil I in your hand so that your thumb is free and stretch your arm straight out. Close one eye and hold the pencil so that the top of the pencil matches the top of your partner's head. Move your thumb down until you see your partner's chin. You now have the length of your partner's head. Move your pencil down so that the top of the pencil is at the chin. (Be careful! Don't move your thumb!) Continue moving your pencil

down counting the number of "heads" that you find when you reach your partners foot. You may have to estimate a fractional part of a "head" for your last measurement. You will probably find that your partner is anywhere from 6 to 8 "heads' tall. Using the same head measurement, find the length of your partner's wrist to fingers measurement, the elbow to finger measurement and the width of the shoulders. Now, switch places and have your partner make the same measurements for you.

Name of Partner		
Height in heads		
Shoulder width in heads		
Wrist to fingers in heads		
Elbow to fingers in heads		

2. Write a ratio in fraction form that compares the length of the head to the total height. Be sure to reduce the ratio if possible. Write ratios for the other measurements comparing each to the total height. Record your results in the table below.

Name of Partner		
Ratio for head		
Ratio for shoulder width		
Ratio for wrist to fingers		
Ratio for elbow to fingers		

3. The ratios that the artist uses are shown below.

Head : Height = 1 : 7

Shoulder Width: Height = 3 : 7

Wrist to fingers: Height = 1 : 7

Elbow to fingers: Height = 2 : 7

Write each of the ratios above as fractions. Compare the fractions for each of your measurements to the fractions for the artist's model. Make a list <u>comparing</u> your measurements to the theoretical measurements using the symbols for greater than, less than and equal to. Hint: To compare fractions, you need to have the same denominators. Show all work and the list in the space below.

4.   Let's compare the ratios that we calculated from the artist's method of measurement to ratios found using a more standard type of measurement. Use a tape measure or a yardstick to find the same measurements that we found earlier. You may find it easier to measure the length of your partner's head and his/her height by marking the top of the head and the chin on a blackboard or white board and measuring the distance between the marks. You can measure from the top of the head measurement to the floor for the height measurement. The other measurements can be found directly. Record all measurements in inches rounding off to the nearest eighth of an inch in the table below.

Name of Partner		
Head		
Height		
Shoulder width		
Wrist to fingers		
Elbow to fingers		

a. Find the ratios between the measurements as you did before using the measurements in inches. Do your work and complete the table below.

Name of Partner		
Ratio for head		
Ratio for shoulder width		
Ratio for wrist to fingers		
Ratio for elbow to fingers		

b. Compare the ratios you calculated when measuring in "heads" to the ratios of the measurements in inches. Were they the same? If they were not the same, why do you think they differ?

c. Use your measurement for the length of your head in inches and the ratios used by an artist for standard proportions (Check part 3) to calculate the height, shoulder width, wrist to finger, and elbow to finger measurements for the "average" person with your size head. Are you "average"?

CIRCULAR RELATIONSHIPS

We use many objects that have a circular shape at home, at school and at work. Three terms can used when describing a circle. The circumference is the distance around the circle; the radius of a circle is the distance from the center of the circle to a point on the circle; the diameter is the length of a line segment that passes through the center of a circle connecting two points on the circle.

You will work with a partner to measure circular objects found in or around your classroom in this activity. You will calculate the ratio of the circumference to the diameter of the circle. Different units of length can be used to measure diameters and radii. We will use both inches and centimeters in this activity.

1. Get a measuring device that will let you measure circular objects in inches and in centimeters.

2. Find 5 - 6 circular objects. These objects may be found in your classroom, in the building or in the area outside of the building. Check with your instructor to find out where you should search for objects to measure.

3. List the name of the object, its circumference (in inches and centimeters), and its diameter (in inches and centimeters) in the table below. Round off the measurements in inches to the nearest eighth of an inch and the measurements in centimeters to the nearest tenth of a centimeter.

Name of Object	Circumference in. cm.		Diameter in. cm.	

4. For each of the objects measured above, <u>write</u> the ratio, $\dfrac{\text{circumference}}{\text{diameter}}$, for each object in the table below. (You may want to change the measurements in inches to a decimal form.) Use your calculator to <u>find the value</u> of the ratio, $\dfrac{\text{circumference}}{\text{diameter}}$, rounded off to the nearest thousandth.

Name of Object	$\dfrac{\text{circumference}}{\text{diameter}}$ (inches)	Result (rounded)	$\dfrac{\text{circumference}}{\text{diameter}}$ (centimeters)	Result (rounded)

How do the numbers in the results column compare?

Are they all approximately the same or are there large variations?

If you have large variations, go back, measure the circumference and the diameter again and recalculate your results.

5. Find the average of the 10-12 ratios that you have in the table.

Does this number look familiar?

What Greek letter is often used to represent this number?

PAINTING THE CLASSROOM

Your classroom must be painted. Your team is responsible for measuring each wall in the classroom as well as any doors, windows, blackboards, etc. The measurements will be used to make scale drawings to help you determine the area that must be painted. The area will be used to decide the amount of paint to buy and the cost of the paint.

1. Measure the length and the height of the first wall using a yardstick or a measuring tape. Round off your measurements to the nearest $\frac{1}{4}$ inch.

2. If you are using graph paper with $\frac{1}{4}$ inch grid marks, let each grid mark represent 1 foot. If you are using another type of graph paper, then decide how many grid marks should represent one foot. Draw a rectangle representing the first wall on the graph paper with the appropriate length and width for the first wall.

3. Go back to the first wall and measure any doors, blackboards, etc. to the nearest $\frac{1}{4}$ inch. Draw these objects on your drawing for the first wall.

4. Use your scale drawing to find the area of the wall that requires paint. Show your calculations for the area in the space below.

5. Repeat 1-4 for each of the other walls in your classroom and show the calculations below.

6. A local building supply store listed interior flat latex paint that covered in one coat for $19.98 per gallon. One gallon covers 350 square feet.

7. Find the total number of square feet that must be painted in your classroom from the calculations in part 4 and 5.

8. Use a <u>proportion</u> to find the number of gallons of paint needed to paint the classroom.

9. Use a proportion to find the cost of the paint using the number of gallons from part 8.

10. A second ad listed a <u>5-gallon container</u> of flat latex paint for $29.90. If this paint also covers 350 square feet per gallon, how much money <u>could be saved</u> by using this paint? Show all calculations.

VOLUME OF FOOD CONTAINERS

You will work with a partner to find the volume of common food containers by measuring the volume and then by calculating the volume. Your instructor will let you know if you should bring containers from home or whether they will be supplied.

CYLINDRICAL OBJECTS

1. To measure the volume of cylindrical cans, fill them to the top with sand, beads or water as instructed. Pour the filling substance into a measuring cup or a volumetric cylinder and measure the volume. Round off as directed. List the name and volume in the table below.

2. To calculate the volume of cylindrical cans, use the formula, $V = \pi r^2 h$. Be sure to measure in the appropriate measurement system, English or metric. Show your work below and list the volume in the table.

Name	Measured Volume	Calculated Volume

RECTANGULAR OBJECTS

1. To measure the volume of rectangular boxes, use sand or beads to fill the box and then measure the volume of the substance with a measuring cup or volumetric cylinder. List the name and the volume in the table below.

2. Use the formula, $V = lwh$, to find the volume of the rectangular boxes. Use the appropriate system of measurement. Show your calculations below and put the results in the table.

Name	Measured Volume	Calculated Volume

COMPARISON OF VOLUMES

1. Compare the measured volumes and the calculated volumes. Do they have the same units?

2. If they do not have the same units, change the measured volume to the same units as used in the calculated volumes. For example, if the volume of a can was found to be 1 cup and the calculated volume was 14 in^3, then you can convert 1 cup to in^3 by using the relationship, 1 cup = 14.4in^3.

The following relationships may be helpful in conversions:
1 cup = 14.4 in^3
1 fluid ounce = 1.8 in^3
1 cm^3 = 1 milliliter

Other volume relationships will be found in your textbook. Show your work below.

Fill in the table below to show the measured and calculated volumes with the same units. Then answer the questions that follow.

Name	Measured Volume	Calculated Volume

3. How closely do the measured volumes agree with the calculated volumes? Be specific. If you have any measurements that are very different, redo that measurement.

4. There are measurements given on the cans or boxes. Does this give the volume of the container?

   If not, what type of measurement does it give?

   Why would this type of measurement be more useful to consumers?

VERIFYING THE FORMULA FOR THE SUM OF ANGLES OF A POLYGON

Your textbook gives the formula for the sum of the measures of the angles of a polygon (S) of n sides to be $S = (n - 2)\, 180°$.

You will work in groups of 2-3 students to collect actual measurements for the angles of polygons with 4, 5, 6, 7, 8, and 9 sides. The points from this data will be plotted and the corresponding line will be drawn. Predictions using this line will be made about the sum of the measures of angles and the number of sides of other polygons. You will compare the results of your measurements to the predicted values from the formula.

Part 1

a. Draw an acute and an obtuse angle in the space below and measure the angle to the nearest degree using your protractor. Record the measure in the blank. (Remember to extend the lines far enough to be able to use the protractor.)

ACUTE ANGLE __________

OBTUSE ANGLE __________

b. Have another member of your team verify your measurements.

Part 2

a. All members of your team should draw each polygon, measure
the angles of the polygon, and find the sum of the angles. Draw
each polygon large enough so you can easily measure the
angles using your protractor. Be sure that you draw polygons
with 4, 5, 6, 7, 8, and 9 sides. Plain paper or graph paper may be
used.

Each member of the team should draw a polygon with a different
shape for a given number of sides. For example, a 4-sided polygon
might have any of the following shapes.

b. Record <u>your</u> data for each polygon in the chart below. Obtain
and record the rest of your team data, find the average sum of
angles for each type of polygon and record. Put the team
member's name in the blank below "sum of angles".

number of sides	sum of angles (your results)	sum of angles	sum of angles	sum of angles	sum of angles (average)

Part 3

a. Graph the data from Part 2 using the number of sides for the horizontal axis and the average sum of the measures of the angles for the vertical axis. Use a scale from 0 to 15 for the horizontal axis. Be sure to extend the vertical axis beyond your actual data so that you can make predictions. Use an entire sheet of graph paper for your graph. Be sure to label your axes, show your scale clearly and plot the points carefully.

b. Draw the best straight line through your data points. (There may be some points that will not fit exactly on your line. Extend the line in both directions.

c. Use your line to find the sum of the measure of the angles of a triangle.

How closely does this value agree with the value that you know?

d. Based on your answer from part c, do you think that your line will give you accurate predictions for polygons with 10, 12, or more sides? Why or why not?

e. Use your line to predict the sum of the measures of the angles for 10-, 12-, and 15-sided polygons. Record your data below in the first two columns.

number of sides (n)	sum of angles from graph (S)	sum of angles from formula (S)

f. Use the formula, $S = (n - 2)180°$, to calculate the sum of the measures of the angles of a 10, 12, and 15 sided polygon and record the results in the chart in e above.

g. How do the two values for S (from the graph and from the formula) compare?

Questions

1.   a. Use your graph to find the number of sides a polygon should have if the sum of the measures of the angles is 1980°.

      b. Use the formula to find the number of sides a polygon should have if the sum of the measures of the angles is 3600°. Show your work.

2.   a. Find the sum of all the angles in the polygon drawn below. Show the size of each angle on the diagram. Hint: all the angles measured must be inside the polygon (interior angles).

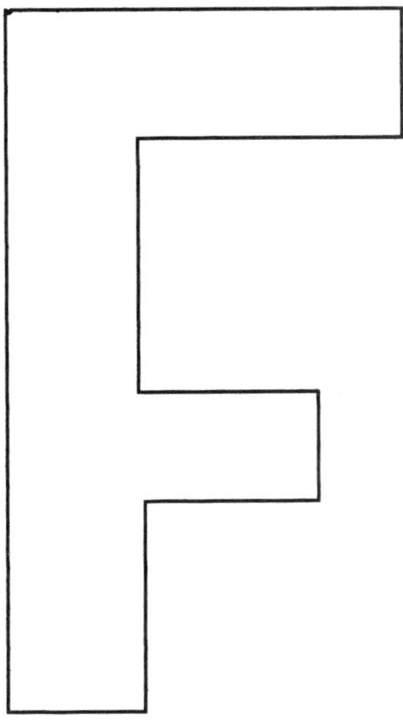

      b. Count the sides and use the formula to find the sum. Your answers should agree. If not, check both parts.

3.   What values of n should not be used in the formula, $S = (n - 2)180°$? Why?